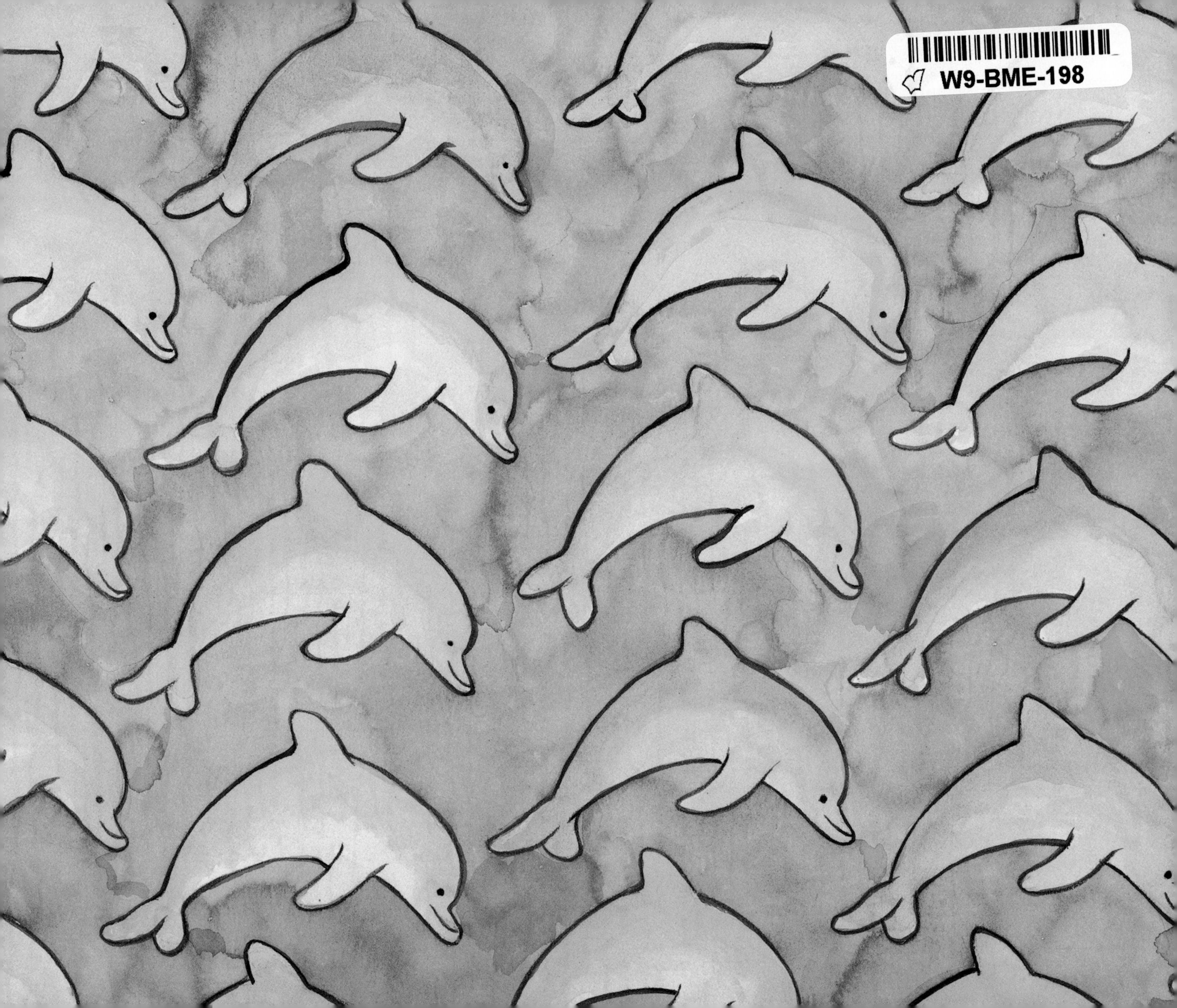

SUCTION
DARTS
chocolate Covered Bananas
3 SHOTS
Win A Dolphin Mascot
HYPNO
ENTER

More or Less

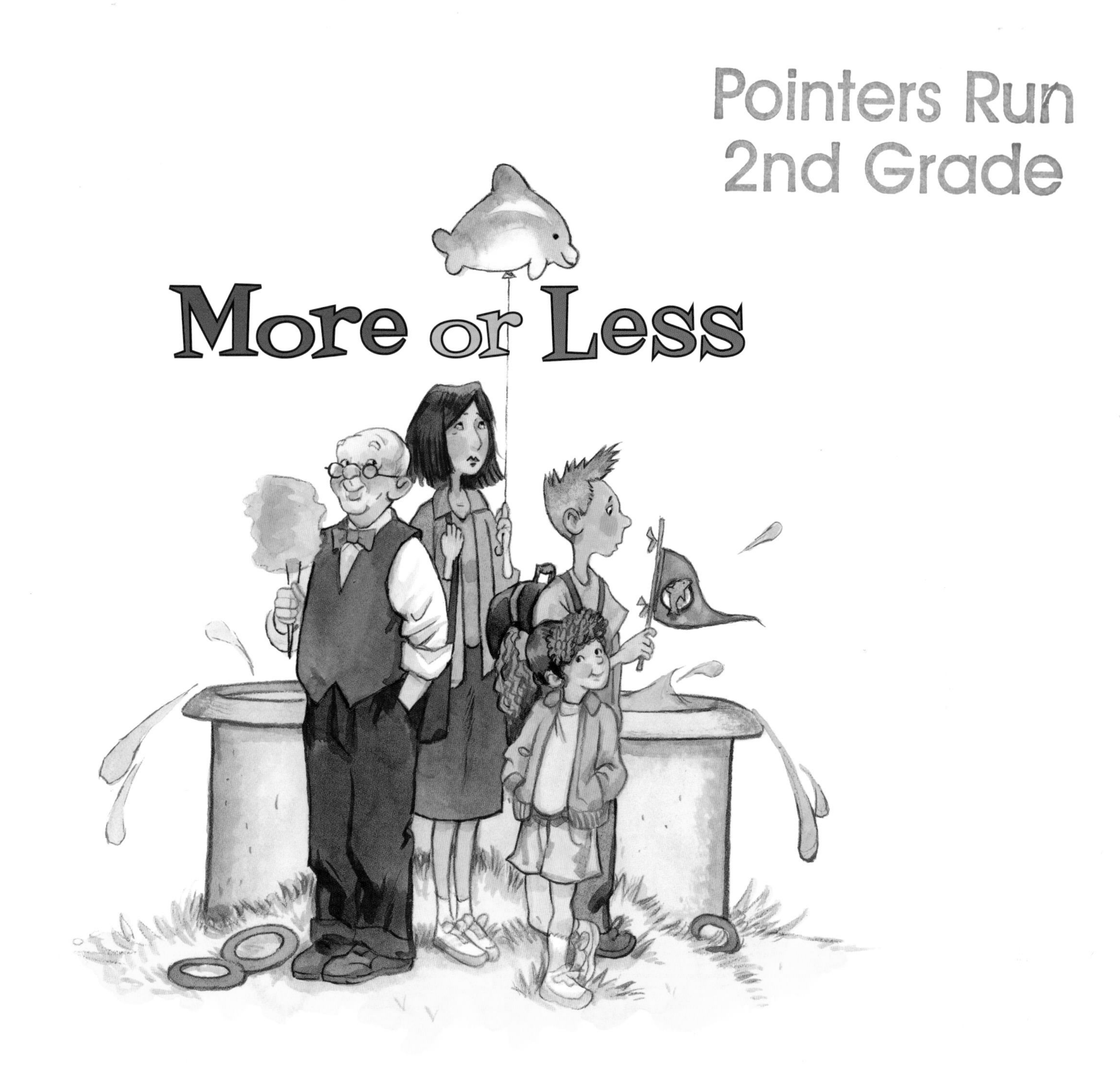

MathStart®

COMPARING NUMBERS

More or Less

by Stuart J. Murphy

illustrated by David T. Wenzel

LEVEL 2

HarperCollins*Publishers*

To Nathan, who is sure to be a master at math
—S.J.M.

To my son, Christopher
—D.T.W.

The publisher and author would like to thank teachers Patricia Chase, Phyllis Goldman, and Patrick Hopfensperger for their help in making the math in MathStart just right for kids.

HarperCollins®, ®, and MathStart® are registered trademarks of HarperCollins Publishers. For more information about the MathStart series, write to HarperCollins Children's Books, 10 East 53rd Street, New York, NY 10022, or visit our website at www.mathstartbooks.com.

Bugs incorporated in the MathStart series design were painted by Jon Buller.

More or Less

Manufactured in China by South China Printing Company Ltd.

Library of Congress Cataloging-in-Publication Data
Murphy, Stuart J.
More or less / by Stuart J. Murphy ; illustrated by David T. Wenzel.— 1st ed.
p. cm. — (MathStart)
Summary: At a school picnic, Eddie uses his knowledge of numbers to outsmart the people who come to his game booth.
ISBN 0-06-053165-7 — ISBN 0-06-053167-3 (pbk.)
1. Arithmetic—Juvenile literature. [1. Arithmetic.] I. Wenzel, David, 1950- ill. II. Title. III. Series.
QA115.M8723 2005 2003027847
513—dc22

Typography by Elynn Cohen 11 12 13 SCP 10 9 8 7 6 ❖ First Edition

Be sure to look for all of these **MathStart** books:

LEVEL 1

Beep Beep, Vroom Vroom!
PATTERNS

The Best Bug Parade
COMPARING SIZES

Bug Dance
DIRECTIONS

Circus Shapes
RECOGNIZING SHAPES

Double the Ducks
DOUBLING NUMBERS

The Greatest Gymnast of All
OPPOSITES

Henry the Fourth
ORDINALS

It's About Time!
HOURS

Missing Mittens
ODD AND EVEN NUMBERS

Monster Musical Chairs
SUBTRACTING ONE

One...Two...Three...Sassafras!
NUMBER ORDER

A Pair of Socks
MATCHING

Rabbit's Pajama Party
SEQUENCING

3 Little Fire Fighters
SORTING

LEVEL 2

Animals on Board
ADDING

Bigger, Better, Best!
AREA

Coyotes All Around
ROUNDING

A Fair Bear Share
REGROUPING

Get Up and Go!
TIME LINES

Give Me Half!
UNDERSTANDING HALVES

Let's Fly a Kite
SYMMETRY

More or Less
COMPARING NUMBERS

Pepper's Journal
CALENDARS

Probably Pistachio
PROBABILITY

Racing Around
PERIMETER

Spunky Monkeys on Parade
COUNTING BY 2s, 3s, AND 4s

The Sundae Scoop
COMBINATIONS

Super Sand Castle Saturday
MEASURING

LEVEL 3

Betcha!
ESTIMATING

Dave's Down-to-Earth Rock Shop
CLASSIFYING

Dinosaur Deals
EQUIVALENT VALUES

Divide and Ride
DIVIDING

Game Time!
TIME

The Grizzly Gazette
PERCENTAGE

Jump, Kangaroo, Jump!
FRACTIONS

Lemonade for Sale
BAR GRAPHS

Less Than Zero
NEGATIVE NUMBERS

The Penny Pot
COUNTING COINS

Polly's Pen Pal
METRICS

Safari Park
FINDING UNKNOWNS

Shark Swimathon
SUBTRACTING TWO-DIGIT NUMBERS

Sluggers' Car Wash
DOLLARS AND CENTS

Mr. Shaw had been the principal of Bayside School for so many years that most people couldn't even remember how long it had been. But now he was retiring.

All the students and teachers, most of the parents, Mr. Shaw's family, and people from the neighborhood had gathered in the school yard for a big picnic in his honor. All around the playground there were booths with games to play.

"Let Eddie Guess Your Age!" was one of the most popular booths. If Eddie could guess a person's age after three questions or less, he won. If it took him four questions or more, the person got a prize. And if he couldn't guess even after six questions, Eddie got dunked in a big tank of water.

Eddie was good—
and lucky. He hadn't
gotten dunked yet.

LET Eddie Guess Your Age!
PRIZES
Dunk Eddie
Rules
3 Questions or less: Eddie Wins!
4 Questions or more: You get a Prize!
More Than 6 Questions: Eddie Gets Dunked!

Clara, one of Eddie's classmates, came along. She spoke in a really squeaky voice so Eddie wouldn't know who she was.

"Bet you can't guess my age," Clara said.

"Is it less than 10?" asked Eddie.

"Yes," said Clara.

"More than 7?"

"Yes."

...5 6 7 8 9 10 11 12...

“So you’re older than 7 and younger than 10. Is your age an even number?” asked Eddie.

“No, it’s not,” squeaked Clara.

“Then you’re 9 years old,” said Eddie. “No prize for you.”

“Aw,” complained Clara in her real voice. “I never win anything.”

“Try another game,” said Eddie. “You never know when your luck will change—Clara!”

One of the parents came over to Eddie's booth. She spoke in a low, growly voice, but Eddie could still tell that she was a grown-up. He thought, *My mom just turned 42. Maybe I'll start there*.

"Are you older than 42?" asked Eddie.

"Yes," muttered the woman.

"Have you had your 46th birthday?" asked Eddie.

"No," she said.

...40
41
42
43
44
45
46
47
48...
Prizes

"Is it an odd number?" asked Eddie. There are two odd numbers between 42 and 46. If the woman said yes, he'd have to ask a fourth question and give a prize away.

"No, it's not," said the woman.

"Then you're 44 years old," said Eddie. "No prize for you."

Meanwhile, Clara had decided to take Eddie's advice.
But her luck hadn't changed yet.

"Sorry," said the woman at the ring-toss booth. "Why don't you try the dart game?"

At Eddie's booth, an older kid came over. He sounded like a teenager, so Eddie asked, "Are you older than 13?"

"Yes," whispered the boy.

"Younger than 15?"

"No," the boy said.

Now I'm in trouble,
thought Eddie. "Older than 20?" he asked.
"No," the boy said.

...12 13 14 15 16 17 18 19 20 21 22...
Rules
3 Questions or less: Eddie Wins!
4 Questions or more: You get a prize!
more than 6 Questions: Eddie Gets Dunke

Between 15 and 20 narrows it down, thought Eddie. "Are you 18?" he asked.

"Nope," the boy said.

"Are you 17?"

"You finally got it," he said. "But it took you five questions."

"Pick a prize," said Eddie.

"Aw, too bad," said the boy. "I wanted to see you get dunked!"

At the booth next door, Clara was still doing her best to win a prize.

"Clara, I don't think this game is for you," said the teacher at the dart game. "Look—isn't that your grandfather over there?"

Clara looked up. She grinned and ran off to Eddie's booth.

The next voice at Eddie's booth sounded really old.
"Are you older than 50?" asked Eddie.
"Yes," said the man.

"Younger than 55?" asked Eddie.
"No," said the man.
"Between 55 and 60?"
"No."
"Younger than 62?"
"No."

. . . 50 51 52 53 54 55 56 57 58 59
Dunk Eddie

All Eddie knew was that this person was at least 62 years old.

"Are you younger than 68?" Eddie asked.

"No," said the man.

Eddie had just one question left before he'd get dunked.

"Are you 69?" he asked.

"No!" said the voice. Eddie heard Clara laugh.

And then . . .

...67 68 69 70 71 72 73 74 75 76 77 78
SPL

Eddie untied his blindfold as he climbed out of the tub. "You must be as old as Mr. Shaw!" he said as he wiped water from his eyes.

"I *am* Mr. Shaw," said Mr. Shaw. "And I'm 70."

Mr. Shaw picked the biggest prize in the entire booth. Then he handed it to Clara.

Clara gave her Bayside dolphin a big hug. "Thanks, Granddaddy!" she said.

FOR ADULTS AND KIDS

In *More or Less*, the math concept is comparing numbers, an important part of understanding the concepts of "greater than" and "less than." *More or Less* also demonstrates the skill of making logical guesses. Rather than making a random guess, children need to learn to look at the available information and ask the questions that will help them to make an "educated" guess.

If you would like to have more fun with the math concepts presented in *More or Less*, here are a few suggestions:

- Read the story with the child. For each character who participates in the guessing game, have the child predict the next question that Eddie will ask before you read on. Stop again after Eddie figures out the correct age and discuss how the questions Eddie asked helped him to arrive at the correct answer.

- Read the story a second time. Stop and have the child come up with different questions that Eddie could ask each person in the story. This will help demonstrate the relationship between the question and the answer, and how important it is to ask the right questions.

- Tell the child you are thinking of a number between 10 and 20, for example. As the child makes guesses, indicate whether each guess is more than or less than the correct answer. Encourage the child to find the number in three guesses. Then have the child think of a number and you make guesses. He or she should tell you if your guess is more than or less than the correct number.

Following are some activities that will help you extend the concepts presented in *More or Less* into a child's everyday life:

Secret Number: Write out clues for a specific number. (For example: "More than 50; less than 60; more than 55; less than 58; an odd number.") Give the child the first two clues and have him or her write down all the possible numbers. One by one, give the other clues. Have the child cross out numbers that are no longer possible until he or she finds the secret number.

Concentration: Make twelve cards, each with a number and the "greater than" or "less than" sign (for example, "< 12" or "> 14"), and twelve cards with only a number. Mix up each set of cards in two separate stacks and turn them facedown. The first player turns up two cards, one from each stack. If the player can arrange them to make a true number sentence (such as 14 < 30), he or she keeps them and goes again. If not, the cards are put back facedown and the next player takes a turn. The player with the most cards at the end wins.

The following stories include some of the same concepts that are presented in *More or Less*:

- ANNO'S MAGIC SEEDS by Mitsumasa Anno
- MATH FOR SMARTY PANTS by Marilyn Burns
- HENRY KEEPS SCORE by Daphne Skinner